Studies in Musical Science & Philosophy
Vol I

The Physics of Piano Unisons

Brian Capleton PhD

Adaptations from published papers

Dr Brian Capleton lectured in Piano Technology at the Royal National College, and is an alumnus of Wolfson College Oxford, the Royal College of Music, Trinity College of Music London, and Dartington College of Arts.

Published by Amarilli Books

Copyright © 2015, Brian Capleton

1st Edition

The right of Brian Capleton to be identified as the
author of this work has been asserted in accordance
with the Copyright, Designs and Patents Act, 1988.

ISBN 978-0-9928141-6-8

A CIP catalogue record for this book is available
from the British Library.

Contents

About this book

This book deals primarily with the physics and mathematics of piano string unison tuning. Readers who want to get a good technical understanding of fine unison tuning, but without too much mathematical detail, and in a more general way that can be related readily to the practical art and experience of tuning, should consult Vol 2 of this series, *Piano Unison Tuning*. For a more comprehensive picture of the whole art of piano tuning, including both practical instruction to advanced level, and accompanying theory, *Theory and Practice of Piano Tuning* can be consulted.

The 'traditional' 19th-century theoretical concepts on which the explanation of piano tuning practice was based, created the first 'traditional theoretical framework' for teaching and understanding the art. However, expert tuners have always been aware of the difference between the early theory's predictions about the kind of acoustical behaviour we might find when tuning the piano, and the kind of behaviour actually encountered.

There has always been a divide between the early theory, and the professional practice by expert aural tuners, which has done little to encourage a wider appreciation of the true complexity of the art in the hands of the expert.

The theoretical understanding of aural piano tuning has of course improved, since the concepts from 19th century theory were first adopted to form the basis of the 'traditional' approach. The first big improvement came in understanding inharmonicity due to string stiffness, a phenomenon more generally known in acoustics as mode frequency dispersion.

The next major leap forward came in understanding how the motions of piano strings are actually coupled through the bridge and soundboard, and how this affects tuning behaviour. It is the latter that this book deals with, in its application to tuning piano unisons.

The general understanding that piano strings can vibrate in two planes rather than just one, and that their motions can be strongly coupled by the bridge and soundboard, gives us a much better insight into the actual, acoustical nature of fine tuning as carried out by expert aural tuners.

The tuning of piano string unisons

The precision tuning of unison string groups in pianos is achieved by piano tuners using a technique that involves listening to the audible 'beats' that can occur when two strings are slightly 'mistuned'.

The rate of beating, or number of beats per second, heard between two 'mistuned' strings, decreases as the two strings are brought into fine tune. 'Traditional' (19th century in origin) piano tuning theory prescribes that the strings of a unison group should be tuned by reducing all beat rates to zero, or in other words, by eliminating the beats.

Artist piano tuners know from experience that there is considerably more to fine tuning unisons than this simple prescription suggests, but the notion of reducing beat rates or attempting to eliminate beating in a unison, is nevertheless basically correct advice as a starting point.

The beats occur when *two strings* are sounding, and the beat rate can be directly and precisely controlled by changing the tension on one string. However, another kind of audible beating often appears in piano tuning, whose behaviour lies largely outside this control. These beats can occur on a *single string* sounding on its own. Such beats are 'traditionally' called *false beats*.

False beats

Every experienced tuner will be familiar with single strings that exhibit 'false' beating when sounding on their own. Such a string would itself be referred to as a *false string*. Almost a century ago in 1907, Jamie Cree Fischer in his book *Piano Tuning / Regulating and Repairing*,[1] spoke of 'false waves' in piano tuning, describing false beats. Ten years later William Braid White referred to 'false beats' in *Piano Tuning and Allied Arts*, believing them to arise from 'segments of the string being strained unevenly, whereby the corresponding harmonics are thrown out of tune'.[2]

White's explanation has possibly led many to confuse falseness with *inharmonicity*, which does, crudely speaking, cause 'harmonics' to be 'thrown out of tune'. Inharmonicity is the slight raising of partials frequencies from what would otherwise be their positions in the harmonic series,[3] and is caused by string stiffness.

In more general terms, inharmonicity is an example of *dispersion* of wave velocities on the string, caused by the fact that the wave velocity becomes a function of wavelength. Inharmonicity is however a feature of string behaviour quite distinct from falseness, and does not itself account for false beats. It is perfectly possible for a string that is not false to exhibit inharmonicity, and it is also theoretically possible for a string to be false, but not inharmonic.

Piano tuners will have encountered false beats, not as some 'strange' phenomenon rarely found, but rather, as an expected, reasonably common occurrence. False beats are not usually welcomed by piano tuners, because they can constitute a hazard or an impediment to the attempt to eliminate beating in unisons and octaves, or to control beat rates in the tempered intervals. Nevertheless, false beats are a part of the natural acoustical behaviour of piano strings *in situ*, and should therefore be embraced by the understanding and practical skills of the artist tuner.

Convention for illustrating decay curves

It is convenient to illustrate audible beats with a visual representation. In general, all the kinds of sounds to which piano tuners listen, can be illustrated as graphs of sound 'amplitude' versus time. For example, the sound of a tuning fork, which begins loudly and ends quietly, might be described as in Fig. I, in which the loudest sound can be seen as the height of the graph at the beginning of the ten seconds, and the quietest sound is at the end, where the height is almost zero.

Fig. I

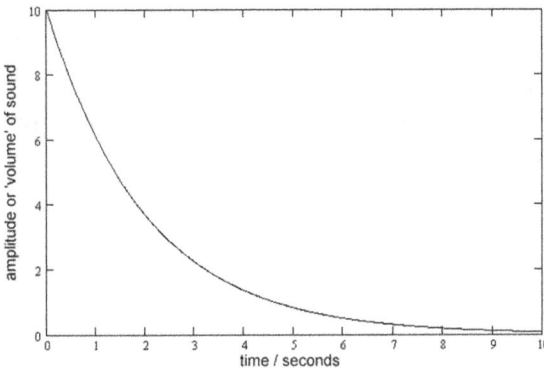

'Decay curve' for the sound of a tuning fork over ten seconds, showing 'exponential decay'

Most sounds encountered in musical acoustics actually begin with a short period of noise known as the *transient*, such as the 'chiff' of a violin bow, the knock of a piano hammer on the string, or the initial noise of the tuning fork as it strikes a surface. For the purposes of these graphical illustrations, the transients of the sounds will be ignored.

Also, there is often a sound at the termination of a tone when the tone is finally damped, such as when the damper of a piano action comes back into contact with the string, stopping its vibrations. This too will be ignored. What is primarily of interest here is the 'decay tone' whose *decay curve* is shown in the graphs.

The curve in Fig. 1 is *exponential*, which means the *rate* at which the volume of the sound diminishes, itself diminishes at time goes on. The sound of the tuning fork diminishes rapidly at the beginning, and slowly towards the end of the ten seconds. This will be familiar from experience.

The period in which the fork sounds loud is quite short, but the period in which it continues very quietly, is relatively long. The tones of piano strings also behave this way, so it is useful to adopt a convention for the graphs, that takes this into account. The convention is to use a *logarithmic* scale on the vertical axis, so that by plotting the data, what would have appeared as an exponential curve, becomes a straight line. This is illustrated in Fig. 2.

Fig. 2

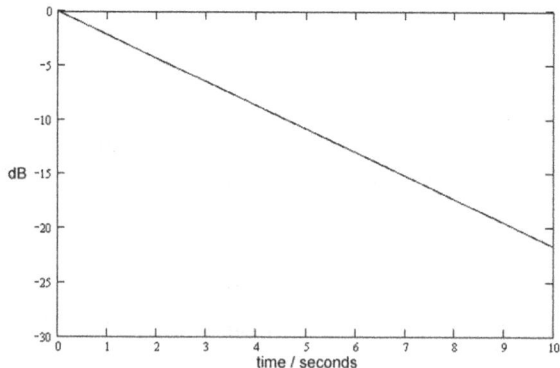

The sound decay curve for the same tuning fork as in Fig. I, using the logarithmic vertical scale, in dB.

Being a straight line, we can now call this decay a *linear* decay, provided we accept as standard practice the semi-logarithmic graphing convention. The terms *linear* and *non linear* are very convenient to use, but relate only to this form of graphing, and are not to be confused with the same terms used in the context of algebra.

The sound level or amplitude now begins at zero, and is measured on a negative scale, in decibels (dB). A drop of 10 decibels corresponds to a drop in sound energy to $1/10^{th}$ of its starting value, but a drop of 30 decibels corresponds to drop in sound energy to $1/1,000^{th}$ of its starting value. A 60 dB drop would correspond to $1/1,000,000^{th}$ of the starting value, and so on. The precise mathematical details need not concern us here.

Beat patterns

The most important non linear decay pattern in piano tuning, is the regular beat, such as might occur at a given partial of a unison string pair, when its two strings are not in proper tune. A regular beat pattern using the dB scale might, for example, appear as in Fig. 3, or as in Fig. 4.

Fig. 3

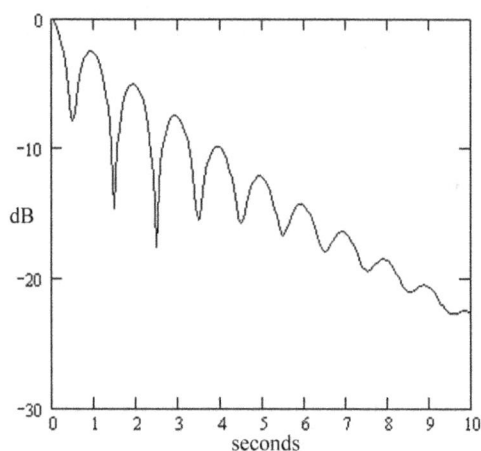

Regular beat pattern with a *beat rate* of 1 per second. The *beat amplitude* of first grows relative to the overall decay, and then decays at a rate greater than the overall decay rate.

Fig. 4

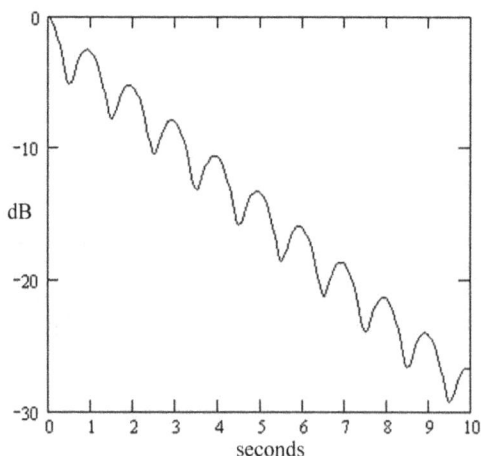

Regular beat pattern with a *beat rate* of 1 per second. The *beat amplitude* stays the same relative to the overall decay rate.

These two figures alone, illustrate how different two beats might be, even though they have the same beat *rate*, or number of beats per second. 'Traditional' piano tuning theory only addresses one aspect of beats - the beat *frequency* or *beat rate* – the number of beats per second.

The two figures illustrate how *beat amplitude* may also be a distinguishing factor. Both decays are non linear, because the decay has a beat pattern, but the overall decay rate in each case, is still linear – a straight line could be drawn along the tops of the beat crests.

Heterodyne beats

'Traditional' piano tuning theory - by which we mean 19th century theory - regards the standard beats whose beat rates are adjusted in the process of piano tuning, as *heterodyne beats*. Audible heterodyne beats are caused by two sound waves with slightly different wavelengths, arriving at the ear simultaneously.

A sound wave in air consists of a moving 'train' of compressions and rarefactions in the air - an alternating sequence of high of and low pressure areas, like even and odd numbered carriages of a passenger train. The wavelength of a sound wave is the distance in the moving 'train' of waves between one high pressure point and the next, or between one low pressure point and the next, which by analogy is like the length of the carriage in the passenger train. High and low pressure points alternate.

Two slightly different frequency sound sources, say, the fundamental tones of two piano strings, will produce sound waves with slightly different wavelengths, but which move through the air at the same speed. At the point where they reach the ear, these two waves will sometimes coincide exactly, with the high and low pressure points of one wave coinciding respectively with the high and low pressure points of the other.

This happens at regular time intervals. In between each such coincidence, the opposite happens - the high pressure points of one wave coincide with the low of the other. This is rather like standing on a station platform as two trains go by at the same speed, but with slightly different length carriages.

As the front of the trains pass, the carriages of the two trains will be side by side, but further down the train, the middles of one train's carriages will coincide with the gaps between the carriages on the other train. This alternating alignment and misalignment in the two sound waves, causes the audible beating.

Coinciding high pressure (or low pressure) areas create a loud sound, whilst if high pressure areas of one wave coincide with low pressure areas of the other, they tend to cancel out, creating quiet sound, or silence. Beats are these alternating loud and quiet in the overall sound.

The phenomenon can be simply illustrated with the schematic diagram, Fig. 5.

Fig 5

C   D   C   D   C

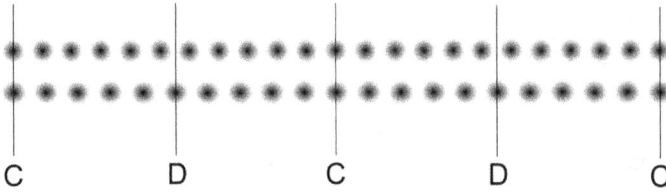

The lines of dots represent high pressure areas in two sound waves consisting of moving 'trains' of high pressure (dots) and low pressure (spaces) areas in the air, both 'trains' moving with the same velocity.

It does not matter whether they are considered moving from right to left or *vice versa*, provided both trains are considered to be moving in the same direction.

The lower 'train' has a slightly longer wavelength – the distance between one dot and the next. Because both 'trains' are moving at the same speed, the positions of the high pressure areas stay the same, relative to each other.

Where two high pressure areas coincide at the lines marked C, they create together a doubly high pressure area overall – an effect called *constructive interference*.

Where a high pressure area coincides with a low pressure area on the other 'train', at the lines marked D, the high and low pressures tend to cancel out, in an effect called *destructive interference*.

The alternating interference effects as the Cs and Ds alternately arrive at the ear, cause the alternating loud (at the Cs) and soft (at the Ds) sound that constitutes the heterodyne *beating*.

This illustration is schematic only – it has only 11 dots of one 'train' to every 12 of the other. A more realistic illustration for two tones around middle C, would have, say, 261 dots on one 'train', to every 262 dots on the other.

Heterodyne beating obeys some very straightforward mathematical rules. These are:

- The number of beats per second (the beat rate or beat frequency) equals the difference in frequency between the two component sound waves. (In Fig. 1, if the distance from one line C to the next, were travelled by the waves in one second, then the upper frequency would be 11 Hz and the lower would be 10 Hz. The difference is 1 Hz, or one beat in one second, a beat wavelength being the distance from one C line to the next.

- When the two waves are 'added', or simultaneously perceived by the ear, the resultant beating sound, will be a single tone whose frequency is the average of the two component frequencies.

Non heterodyne beats

A little thought should suggest that this is not the only way to produce a beating effect in a sound. A single sound can have its amplitude modulated in a beat pattern, at source - for example, if a recorder or flute note is played with vibrato. This too would be beating, but it would not be heterodyne beating.

On the other hand, two recorders or flutes, playing the same note without vibrato, but slightly out of tune with each other, would cause heterodyne beating to be heard. 'Traditional' (19th century based) piano tuning theory treats a pair of sounding piano strings in much the same way as the latter example of the 'out of tune' recorders or flutes. Contemporary acoustical theory treats the piano string pair rather differently.

In the 'traditional' theory, the physics assumed would be essentially correct for two separate, non beating sound waves that reach the ear, and then produce heterodyne beating as an interference pattern between themselves. It is actually still this kind of physics that is assumed in 'traditional' tuning theory, even if the 'mixing' of the two slightly 'out of tune' frequencies is considered to take place in the string-bridge-soundboard system, rather than in the air.

In either case, the assumed physics 'works' in essentially the same way. In contemporary acoustical theory, on the other hand, not only will a beat pattern be already present in the soundboard system, before radiation to the air, but each of the two frequencies responsible for the beat are not those separately associated with each separate, individual string. They are not, as it were, individual string frequencies, but rather, are two frequencies of another kind, that arise within the complete two-string system.

The 'mixing' of these different frequencies in contemporary theory, allows that for larger 'mistunings', the strings more or less obey those straightforward mathematical 'heterodyne-like' rules cited above, but for finer 'mistunings' inside certain limits, quite different behaviour can take place.

The behaviour is described in contemporary theory by the physics of a system comprised of strings that are coupled by the soundboard and bridge, rather than the mechanics of a system comprised - as in 'traditional' theory - of two separate but simultaneously sounding strings.

One important difference between the contemporary physics approach and the 'traditional' mechanics approach, is that contemporary theory can demonstrate from a more generalised point of view, what kind of behaviour must take place under certain generalised conditions, by considering a system as a whole. To those familiar only with the 'traditional' (19th century) theory of the mechanics of beats, the contemporary understanding provided by the modern approach of physics, will hold some surprises.

Normal modes

To describe the physics of false beats, it is necessary to deal with one of the most important concepts in contemporary acoustics - normal modes. The motion of an acoustical system can be very complicated, and one of the standard practices in modern acoustics is to analyse such complicated motion through the principle of superposition. The technique involves first looking for possible modes of motion in which all parts of the system are vibrating with the same simple harmonic frequency.

Simple harmonic motion is motion, for example, like that of a tuning fork or a simple pendulum. (Mathematically, the acceleration is proportional to the distance from the position of equilibrium).

Any such mode of motion is called a normal mode, and any possible, more complex general motions of the system then occur as the superposition or adding together of the 'simple' normal modes. The word 'normal' in the term 'normal mode', essentially means that any normal mode can exist by itself without causing any other normal mode to be brought into motion.

A given acoustical system will exhibit different general modes of motion that are more complicated, depending on how it is started into motion in the first instance.

Normal modes, as we said, are generally simple harmonic modes of motion (but it is possible to have a system where this is not the case). This means a normal mode is generally sinusoidal motion of just one frequency ('degenerate modes' can occur with the same frequency), which in a 'musical instrument system' will radiate a 'simple tone' rather like the tone of a tuning fork.

In practice, a general motion producing a musical tone that decays, like the tone of a piano string, can still be effectively analysed into simple harmonic normal modes. Mathematically, this can be handled by using complex frequencies, whose real parts determine the number of cycles per second, and whose imaginary parts determine the decay.[4]

In piano tuning, it is recognised that the tone of a piano string can be analysed into a set of component partials, sometimes called 'harmonics' or 'overtones' by piano tuners. The frequencies of these partials are arranged approximately in the well known harmonic series.

These partials can actually be heard as discreet simple tones of different pitches, within the overall tone of the piano string. Standard text books describe these audible partials as each being 'mapped' from a single normal mode of the vibrating string, which is a standing wave[5] of a specific wavelength that can exist on the string.

A decaying normal mode of a piano string would be expected to have a linear decay on the semi-logarithmic graph. The idea that an audible partial is 'mapped' from a single normal mode breaks down when we encounter false partials, precisely because false partials beat, rather than decaying linearly.

Superposition of normal modes

False beats appearing as a regular beat pattern in an audible partial, suggest that the partial is 'mapped' from two superposed normal modes. In an acoustical system we can add together (superpose) various normal modes in various proportions, to get various modes of general motion.

If we were to superpose two normal modes that happened to be close in frequency, we would expect them to produce a general motion mode that exhibited a beat pattern in some way. However, none of the normal mode frequencies in a harmonic series or a series similar to it, are close enough to produce a significant beat pattern on superposition.

In the case of a piano string vibrating in one plane, the frequencies of the normal modes will occur in a set corresponding reasonably closely to a harmonic series of frequencies.

If we were to superpose any two adjacent normal modes (i.e. modes 1 and 2 or modes 9 and 10) from a harmonic series, the beat rate between these modes turns out to be equal to the fundamental[6] frequency (since the beat rate is equal to the difference between the mode frequencies, and the difference between two adjacent frequencies in a harmonic series is always equal to the fundamental frequency).

So, for middle C40 this would be around 262 beats per second, and even for bottom A1 it would be more than 27 beats per second. Such beat rates are too fast to be perceivable as such.

The existence of an audible, false partial, whose beat rate may be, perhaps, in the order of 2 or 3 beats per second, therefore indicates the presence of normal modes whose frequencies are much closer than those found in the harmonic series associated with the piano string.

The two-plane single-string model

The 'traditional' theoretical model for the piano string treats the transverse motion of the string in one plane only, and this model precludes the possibility of normal modes sufficiently close in frequency. Normal modes sufficiently close in frequency can be found in the theoretical model of a single string, when motion in two planes is accounted for.

A real piano string has two boundaries to the speaking length – the top bridge, *capo d'astro* bar or agraffe at one end, and a soundboard-bridge (long bridge or bass bridge) at the other. The acoustical function of the first, rigid boundary, is to cause the travelling waves that are responsible for the creation of the standing waves, to be reflected at the boundary, back along the string's speaking length.

Waves are partially reflected at the soundboard-bridge also, but this boundary also has the important function of transmitting energy from the string to the soundboard, and hence to the surrounding air. In order to achieve this, the soundboard-bridge is capable of (small) vibratory motion.

Mathematical and physical description of motion in two planes

In a single plane, small amplitude normal mode motion in y of an ideal string, tensioned between rigid boundaries, with speaking length along the x axis, can be written in the form

$$y_n(x,t) = \text{Re}\left[C_n \exp(in\omega_0 t)\sin\left(\frac{\pi n x}{l}\right)\right]$$

$$n = 1, 2 \dots \infty$$

(1)

where $\omega_0 = \frac{\pi c}{l}$ is the fundamental angular frequency, c is the phase velocity and l is the speaking length. C_n can be regarded here as containing both amplitude and phase information.

The series for n constitutes the mode numbers of modes with the harmonic series of frequencies $n\omega_0 = \omega_n$, determined by the phase velocity and the speaking length. In the case of a real piano string the mode frequencies would be subject to dispersion due to string stiffness (inharmonicity), but this will be of no consequence in the following discussion.

Replacing one of the rigid boundaries by a bridge boundary capable of small motion, results in a small change to the frequency ω_n, given by[7]

$$\delta\omega_n = iZ_n Y_n \omega_n / \pi$$

(2)

where ω_n is the frequency for a rigid boundary, Z_n is the wave impedance of the string, and Y_n is the complex bridge admittance.

What happens if both the string and the boundary are capable of motion in *two* transverse planes? To begin with, the admittance of the bridge may depend on the transverse direction in which it is measured. The admittance Y_n in Eq. (2) must reflect this fact. Also, if the wave impedance of the string, Z_n, depended on the transverse direction of the wave on the string, then Z_n would have to reflect this fact also. Such a situation could occur, for example, with a non uniform string.

The most general method of writing these quantities when they are dependant on transverse direction, is to specify them not as a scalar quantity, but as a 2X2 (diagonal) matrix. If either or both are a matrix, then the change to the frequency $\delta\omega$ must itself be written as a 2X2 matrix. The resulting two eigenfrequencies (a 2X2 matrix must have two eigenvalues) may or may not be equal, depending on whether the bridge admittance is isotropic around x.

If the bridge admittance is isotropic its admittance matrix will have equal eigenvalues and the resulting eigenfrequencies for ω_n will be equal. If the bridge admittance is anisotropic then the matrix eigenvalues will differ, in their real parts, imaginary parts, or both. If they differ in their real parts there will be two normal modes with slightly different frequencies, replacing the one ω_n that would be present in the one plane system.

It is therefore convenient to define the *partial number* $j = 1, 2 \ldots \infty$ as corresponding to the series of n values, but labelling motion that may contain two eigenfrequencies (in place of a single frequency ω_n) as a result of transverse anisotropy in the reactive part of the bridge admittance. These will be close compared to the separation of frequencies between two adjacent members of the harmonic series (in which the ω_n values fall), and can result in an audible false beat at partial j.

It is important to appreciate that the system is not one in which we can simply say that the string in a grand piano, for example, vibrates at different frequencies in the vertical and horizontal planes. If this were so, then we would in any case not expect the vertical hammer strike to initiate any horizontal motion.

Rather, in the general mathematical description, where there is two-dimensional transverse motion of string and bridge, and a bridge that can have a resistive component, individual normal modes are generally considered as *elliptically polarized*. In elliptical polarization a point on the string will typically prescribe an ellipse in a transverse plane at right angles to the string's length, rather than a straight line, as in the one plane model.

Circular and linear polarizations, where the point moves in a circle, or in a straight line as in the one plane model, can still occur but are just special cases of elliptical polarization and thus still fall within the general description for elliptically polarized modes.

However, in the case where the reactive part of the bridge admittance is also anisotropic, the two eigenfrequencies of partial j will differ in their real parts, and two modes in place of the one ω_n will emerge, at two different, but close frequencies. In general, the transverse motion of the bridge under these conditions will describe a *parametric locus*. This is a path with co-ordinates that are functions of a single parameter, in this case, time.

Such paths are best known as *Lissajous figures*. In the following section we can see how, in the case of two eigenfrequencies whose difference is very small compared to the frequencies themselves, the motion will alternate between two almost linearly polarized motions in different directions, the path in between consisting of a series of almost elliptical loops of increasing and then decreasing diameter. The oscillation from one pole to the other takes place as a result of the cyclically varying phase difference between the two fixed, but different eigenfrequencies.

Schematic illustration of bridge motion

The general manner in which a false beat can occur when the bridge boundary is anisotropic in its reactive part, can be illustrated through a simple sequence of Lissajous figures.

Energy lost at the boundary, from the string to the bridge-soundboard structure, must be accounted for in two transverse directions (even if loss in one direction were to be zero). The two energy loss axes can be regarded as orthogonal 'antenna' directions, which in the case of a grand piano, will be in the vertical and horizontal. Where the reactive part of the bridge admittance is anisotropic, general motion at partial j can be resolved into two linearly polarized modes not necessarily aligned with these antennae.

The theoretical transverse motion of a point at the bridge can be represented graphically, by using the polarization axes as the graph axes. The following Lissajous figures show the developing transverse path of a point representing the bridge boundary of the string, when the component frequencies in the y and z (the x axis being along the string length) directions differ. In theory, the polarizations may be inclined at any angle to the horizontal and vertical axes.

Here, they are placed vertical and horizontal, since the loci are functions of values in these co-ordinates, and the path is expressed as a conventional graph with vertical and horizontal axes. For clarity the maximum 45 degree angle of strike axis to the normal modes is adopted. To represent a situation in a grand piano, this model simply needs to be rotated by 45 degrees, so that the strike axis is vertical, and the normal mode co-ordinate axes are at 45 degrees to the vertical and horizontal.

For visual clarity the frequencies are now set to just 1 Hz and 1.1 Hz, which reduces the number of loops, and the initial amplitudes of the modes in each co-ordinate are set equal. The precise pattern formed by any locus will of course depend on the frequency difference between the modes, the relative initial phases and displacements, and decays.

For clarity, in this example the decay is set as negligible in one beat cycle, and the other parameters are arranged so that the locus pattern is symmetrical – half way through the beat cycle the path retraces itself *exactly* back to the starting position, ready for a new cycle.

Fig 6 (a)

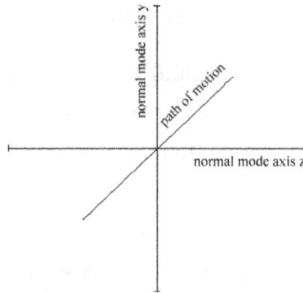

(a): The frequencies in the y and z axes are the same, so the motions in y and z remain in phase, and the locus is a straight line at 45 degrees to the mode axes.

(b)

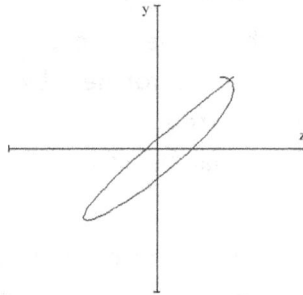

(b): The frequencies are 1 Hz in y and 1.1 Hz in z – the motions in each axis start to become out of phase. This is the path covered at 1 second.

(c)

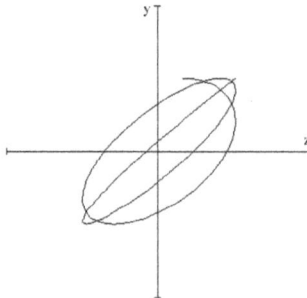

(c): The path at 2 seconds.

(d)

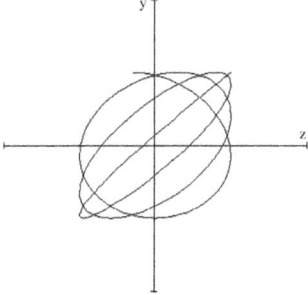

(d): Path at three seconds.

(e)

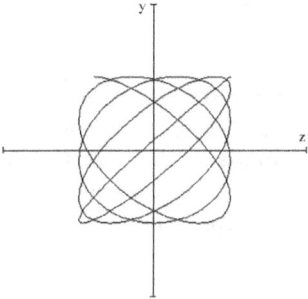

(e): Path at four seconds.

(f)

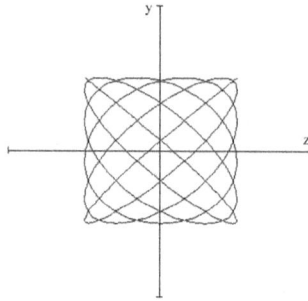

(f): Path at five seconds. The pattern is here complete – the point subsequently retraces the existing locus in the reverse direction for the next five seconds.

Fig 15 (a – f): Path of a point schematically representing the moving bridge boundary of the string, for one partial *j*, when motion begins with equal displacements in the y and z axes. To more literally represent the situation in a grand piano, the graph would have to be rotated by 45 degrees.

Fig 6 (f) shows the path at $t = 5$ seconds. The path starts at the top right corner, with equal displacements in z and y, and finishes at the top left corner of the pattern. The point would subsequently retrace the same locus backwards to the starting point. The insertion of antennae axes at a full 45 degrees inclination (for clarity) to the mode axes (Fig 14), allows the nature of the resulting beat to be seen.

Fig 7

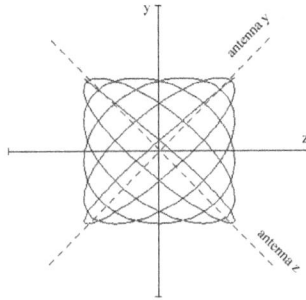

Path of the point Fig 6 (f) shown
relative to antennae axis directions

It can be seen that the motion begins with large amplitude in the y antenna axis, and small in the z antenna axis. The amplitude in the y antenna axis subsequently reduces as the amplitude in the z antenna axis increases. Finally, at 5 seconds, the initial situation is fully reversed – the amplitude in the z axis is the large amplitude, while the amplitude in the y axis is small.

The amplitude in the mode axes themselves, remains unaffected. In this illustration it is not only unaffected, but unlike the case in a real piano, it is also constant, because there is no decay. After 5 seconds the path is retraced in reverse, so the amplitude in z decreases as the amplitude in y increases, until at 10 seconds the initial conditions are recreated.

The motion thus oscillates between motion in the y antenna plane and motion in the x antenna plane. Similar motion can be expected in the transverse displacement of the string along its whole length. For a given partial j, the string in a grand piano would vibrate first in the vertical 'strike' direction perpendicular to the soundboard, and then horizontal motion will increase as vertical motion decreases. After the horizontal motion reaches a maximum, it will start to reduce as vertical motion starts to increase. The string vibration thus oscillates between motion in two planes, with quasi-elliptical motion taking place in between. This kind of motion has been observed in bass strings by Tanaka, Nagai and Mizutani.[8]

Cyclical variation of partial frequency

Antenna y will thus 'transmit' a frequency with a beat modulated amplitude, and so will antenna z, but the beat patterns will be 180 degrees out of phase with each other.

The frequency of motion in the y mode axis here is 0.1 Hz slower then the frequency in the z mode axis. When an antenna axis is aligned somewhere between the mode axes, the motion in the antenna axis direction will approximate to a simple harmonic frequency somewhere between the two extremes 1 Hz and 1.1 Hz in the mode axis directions.

It is important to note that empirical measurements of frequency in the vertical and horizontal directions may show no difference, when in fact two different eigenfrequencies are present but parametric motion resolves to axes at 45 degrees to the vertical and horizontal.

Also, if these axes are not at exactly 45 degrees to the antennae axes, then beating in one antenna can theoretically occur in a 'signal' with a different frequency to the other 'signal' in the other antenna, resulting in a radiated beat pattern whose alternate beats are *at two different pitches*, depending on the parameters.

Both the 'standard' beats that tuners adjust in piano tuning, and false beats, must arise from superpositions of normal mode motions being mapped to one audible partial. The false beat however, may differ from the 'standard' beat, in that its frequency can vary cyclically with the beat itself.

The relationship between frequency variation and false beat rate can be considered as a potentially important indicator of single string behaviour in empirical research. The fact that a false beat can have both cyclically varying amplitude (the beat) *and* cyclically varying frequency, also has important consequences in practical piano tuning procedures.

Where a single string is not false in a partial *j*, the motion in one antenna alone would be expected to decay linearly. Linear single string decays were reported by Hundley, Benioff and Martin, who concluded that the vibration of a single string at a particular frequency 'does not typically exhibit a multiple decay rate'.[9]

However, recordings made by Weinreich in 1977 of a single string through a microphone placed near the piano,[10] revealed a dual decay rate. It is sensible to conclude that single strings may or may not decay linearly, and that even if a partial is not false, it may exhibit a linear or a dual decay rate. Weinreich's model for piano string unisons coupled by the bridge will form the basis of the following descriptions of unison behaviors.

Such variable results would be dependent on the possible differences in parameters. Specifically, a dual decay rate in the single string would require a superposition of decays in *two* mode axes.

Physical bridge properties – the bridge pin

One might have expected, from the soundboard structure and design, that the reactive part of the bridge admittance should be much greater in the strike direction than in the direction parallel to the soundboard surface. Weinreich's measurements show, on the contrary, that the reactive part is approximately the same for both directions, and is only of comparable magnitude to the resistive part, in the vertical direction.[11] Experimental results conclude that the angular variation of the reactive part of the bridge admittance is at least a factor of ten smaller than the variation in the resistive part.[12]

A possible explanation for this is that the bridge pin (side draft pin) at the end of the speaking length, plays a major part in determining the admittance at the string boundary, in the direction parallel to the soundboard surface.

In effect, the bridge pin may considerable lower the reactive admittance in this direction, making it almost isotropic. This role of the bridge pin would be consistent with its capability to also cause localized anisotropy in the admittance.

Fine tuning 'pure' unisons

Normal modes

Two piano strings that are to be tuned to unison are part of a system that is comprised of the two strings that will be sounding, plus the entire string-bridge-soundboard structure. In particular, the two strings are *coupled* by the bridge.

The bridge coupling means that the system does not generally behave as the prescriptions of 'traditional' piano tuning theory suggest. When the two strings are sufficiently 'mistuned', the unison at each audible partial *beats*, as 'traditional' theory states.

The beating in the audible partial arises because the partial is 'mapped' from two normal modes of slightly different frequencies. However, in a system comprised of two piano strings attached to common soundboard or bridge, these two modes are not normal modes of the individual strings, as 'traditional' theory presupposes.

They are normal modes of the complete coupled system consisting of the two strings plus the bridge (whose properties are determined by the whole string-bridge-soundboard structure).

The normal modes of a coupled system are typically different from those of the uncoupled, individual oscillators that make up the system. For example, a single pendulum (in one plane) has a single normal mode, consisting of its swinging back and forth.

A system comprised of two identical pendulums coupled by a weak spring, however, will have *two* normal modes. One normal mode consists of both pendulums swinging back and forth in opposite directions, so that the spring alternately extends and compresses, Fig. 8:

Fig. 8

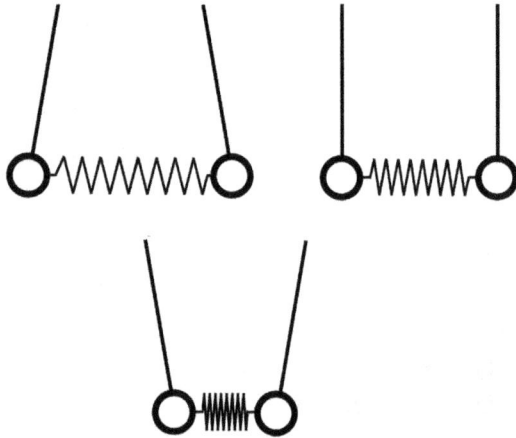

The other normal mode consists of both pendulums swinging in the same direction, so that the spring remains the same length throughout, Fig. 9:

Fig. 9

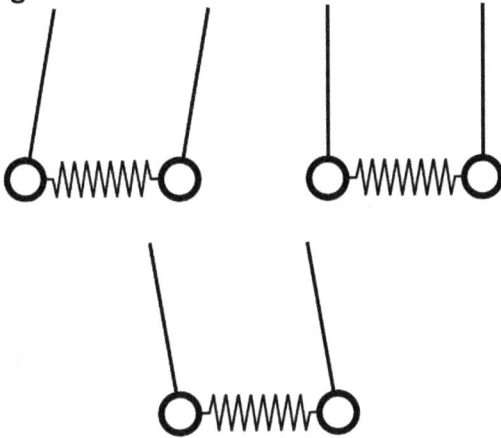

Similarly, in the case of a unison pair of piano strings vibrating in one plane, the normal modes of each isolated string are the standing waves on the string. However, when *both* strings are sounding, coupled by the bridge, there are in the coupled system *two* normal modes replacing each one of the 'single string' normal modes previously.

One mode is *symmetric*, in which the standing waves on each string are moving transversely in the same direction, *in phase* with each other. The other mode is *antisymmetric*, in which the standing waves are moving transversely in opposite directions, *out of phase* with each other. *General motion* of the coupled system, when both strings are sounding, will then be a superposition of these symmetric and antisymmetric modes, each individual mode involving *both* strings.

In the case of the pendulums, the 'antisymmetric' mode in which the pendulums are in opposite motion, alternately 'squashing' and 'stretching' the coupling spring, will be a 'faster' mode, with a higher frequency, then the 'symmetric' mode in which the coupling spring remains the same throughout. Each uncoupled pendulum on its own, would nevertheless have the same frequency.

Similarly, in the case of the piano strings, for any given partial number, the symmetric and antisymmetric modes may have different frequencies, even though both strings are tuned so they have identical partial frequencies when sounding separately.

The meaning of *mistuning* and *mode coupling*

The situation for piano strings, however, can also be reversed. With certain parameters, at any partial number, the two strings may have different frequencies when sounded separately, but when sounded together, there may be no beat.

This situation arises because the coupling bridge - unlike the spring between the two pendulums - has a 'built in' dissipative component designed to transmit energy away from the strings, for amplification by the soundboard and radiation to the air.

For a given partial number, dissipative coupling can cause the frequencies of the symmetric and antisymmetric modes of the coupled system, to be closer than the mode frequencies of the uncoupled strings. It is the difference between the mode frequencies of the *uncoupled* strings that is referred to in the physics as 'mistuning'.

The 'drawing together' of the frequencies of the symmetric and antisymmetric modes of the *complete coupled system*, is referred to as *mode coupling*.

The Weinreich model – for fine tuning

In the Weinreich model two string of a unison are considered coupled by the bridge. The bridge may have *resistive* and *reactive* properties, which are different in character. Its resistive or 'friction-like' properties are the mechanism that transmit energy away from the string for radiation by the soundboard into the air.

The reactive properties are essentially inertial and elastic – they are 'spring-like' and 'mass-like' properties. All these properties are quantified in what is measured or defined as the bridge *admittance*.

For string motion in one plane, the unison at one partial is represented in the Weinreich model by a 2 X 2 *dynamical matrix* derived from perturbation theory. The size of the matrix comes from the fact that there are two strings in the system, both vibrating in the same plane.

Matrices are a powerful tool used in contemporary acoustics (and the other sciences) to handle complex systems, in this case, a number of oscillators joined through a coupling that has both reactive and resistive components. Matrices are essentially tables of values or variables arranged in rows and columns.

In the following discussion the term *eigenvalues* refers to a set of solution values for a given matrix, and an *eigenfrequency* which is computed using matrices, can be thought of as representing the frequency of a normal mode.

A normal mode roughly translates to what piano tuners call a partial. However, one audible partial may contain more than one normal mode. It is when this happens, that we might hear what tuners call a false beat.

The behaviour of the unison as a function of mistuning between the strings can be predicted from the matrix eigenvalues. The model can be used to show the unison behaviour for various bridge types, ranging from (hypothetically) purely resistive to a ('realistic') bridge that is both resistive and reactive.

Actual, experimental measurements by Weinreich and others suggest that in the case of a real piano bridge the resistive and reactive properties of the bridge in the strike direction are about equal. So a realistic bridge represented in the theory, should be about equally resistive and reactive.

Essentially, the Weinreich model shows that tuning the unison is not just a question of beats or no beats, even for a single partial. We can regard 'mistuning' as the frequency difference between two partials with the same partial number, when measured with the strings sounding individually, in isolation from the other string.

'Traditional' theory states that the resultant beat rate will be equal to the mistuning. In fact, this is only true for relatively large mistunings. For very small mistunings of the kind adjusted by an expert aural tuner in the finest tuning, this rule breaks down. Rather, within a *range* of mistunings, the beat will be slower than 'traditional' theory supposes, or in some cases may vanish altogether, even when mistuning is still present.

At the same time, where 'traditional' theory supposes the beat rate will be zero at zero mistuning, the Weinreich model also shows that there may *never* be a truly zero beat rate at zero mistuning.

The Weinreich model shows that inside the finest tuning range, the beat rate may remain approximately the same, whatever the mistuning, and that inside this range it is the partial *decay rate*, rather than the beat rate, that is most critically dependant on mistuning.

The decay rates of the partials are important in determining the tone of the unison. Weinreich stressed that tonal differences from note to note caused by 'hammer irregularities' (or other factors), can therefore be counteracted through the control of mistuning *inside* this fine tuning range.

When the expert tuner tunes a unison, it is usually the high frequency end of the spectrum that receives the finest tuning. In musical and hearing terms, the partials in the higher spectrum form a high frequency 'microtonal cluster'.

It can be speculated that if the decay rates of these high partials are relatively fast (due to additional internal friction and air damping) then they may fall relatively easily inside the range of mistuning necessary for strongly coupled behaviour. However, even in a hypothetical complete absence of beating, any unequal decay rates in the partials that make up this microtone cluster, will inevitably create audible 'movement' in the tonal quality of the upper spectrum.

This 'movement' is theoretically not true beating of any kind, but being audible 'movement' the tuner may rightly aim to eliminate it in the same way. Thus, in the finest tuning that aims to eliminate this movement, it may still be that it is in fact decay rates, rather than beat rates, that are being adjusted by the tuner.

Interpreting the Weinreich model

The Weinreich model's implications for piano unison tuning can sometimes be misrepresented or misunderstood.

Perhaps the most striking part of Weinreich's paper *Coupled piano strings* is the first set of results which are for a conjectured, purely resistive (dissipative) bridge. For such a bridge, the coupled system's mode frequencies do not so much 'couple' as dramatically *lock together* over a relatively large coupling region, or range of mistuning.

This can be seen in the graphs of the eigenfrequencies for the coupled system of two strings plus the bridge, at one partial number, Figs. 10 and 11.

The model for a purely resistive bridge illustrates exceptionally clearly the physics under these parameters, but it should be noted that Weinreich himself continued by verifying that the bridge must be considered as both resistive *and* reactive.

The reactive component of the bridge in fact 'works against' the resistive component's tendency to 'pull together' the coupled system's mode frequencies. The resistive component acts to reduce beat rates to less than what would be expected from 'traditional' tuning theory, whilst the reactive serves to counteract this reduction.

Fig 10

A

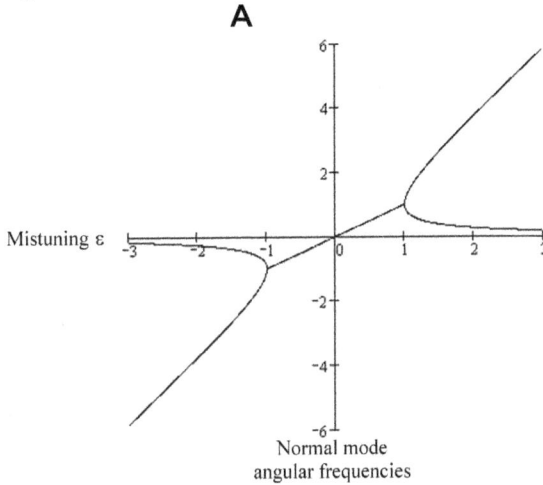

Mistuning ε

Normal mode
angular frequencies

The real eigenfrequencies of Weinreich's matrix for a purely resistive bridge. The mistuning units are equal to the single string decay rate. For mistunings smaller than ± 1, the two eigenfrequencies 'lock together' at one frequency. In the mid range of the compass on a small piano, this might occur when the mistuning is equivalent to what would be expected from 'traditional' theory to give a beat rate of about 0.5 Hz.

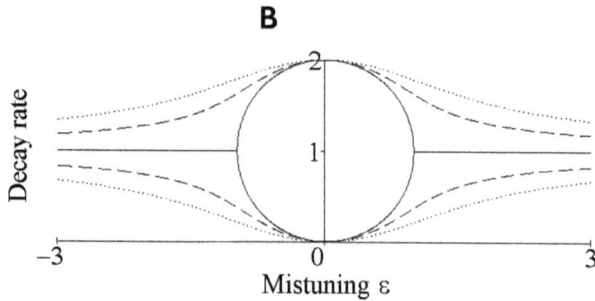

The imaginary eigenfrequencies (mode decay rates) of Weinreich's matrix for various bridges. Solid line = purely resistive bridge, dashed and dotted = bridges with both resistive and reactive components. For mistunings greater than ± 1, there is only one decay rate, equal to the single string decay rate. For mistunings smaller than ± 1, there are two decay rates, the lower one for the antisymmetric mode and the upper one for the symmetric mode.

At zero mistuning the antisymmetric mode has zero decay rate because both strings are acting equally and oppositely (exactly out of phase) on the bridge, which means the bridge does not move in response to this partial, and becomes in effect, rigid. In the symmetric mode at zero mistuning, both string motions are in exactly phase, causing twice as much force on the bridge as the single string, and doubling the single string decay rate.

Fig 11

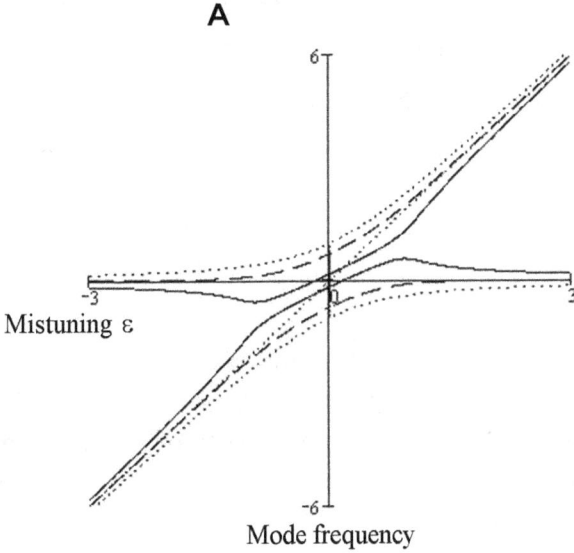

The real eigenfrequencies of Weinreich's matrix for various bridges that are both resistive and reactive, with various resistive and reactive proportions. The solid line has the greatest resistive proportion. The mode frequencies *never* coalesce – but they are drawn initially closer in the 'coupling region', as mistuning decreases. However, as mistuning approaches zero, the eigenfrequencies are *further apart* than they would have been in the absence of bridge coupling.

Within the range of ± 1 mistuning the beat rate (as the difference between the two mode frequencies) is not strongly affected by the mistuning (the lines are roughly parallel). Without coupling, the eigenfrequencies follow the horizontal axis and the dotted diagonal straight line through the origin.

47

Purely resistive bridge

Looking at Fig. 10 A, it can be seen that for mistunings less then ± 1 (1 being equal to the single string decay rate), the two mode frequencies 'lock together'.

A beat can only be present where there are two different mode frequencies, so within the whole of this mistuning range, there is no beat. In the middle compass, ± 1 mistuning might typically be equivalent to a mistuning of around 0.3 or 0.5 beats (per second) in the fundamental, by 'traditional' theory.

Outside this range the mode frequencies separate and approach what they would be in 'traditional' theory. Thus we would find that in the process of slowing the beat rate in fine tuning, the beat in the fundamental would be reduced to around 0.3 or 0.5 beats, but then it would rapidly disappear, leaving the unison beatless, despite small changes of tension in the string.

Looking at the solid line in Fig. 10 B, it can be seen that there are *two* mode decay rates inside the mistuning range ± 1. This means that inside the mistuning range ± 1 there will be a *dual decay rate*. At zero mistuning the two normal modes are distinct.

They are (1) a *symmetric* mode in which the strings are moving 'in phase' with each other, and (2) an *antisymmetric mode* in which the two strings are moving in opposition 'out of phase' to each other. In the antisymmetric mode the forces the two strings exert of the bridge cancel out with the result that the bridge does not move and there is no decay.

This is the lower of the two solid lines. The symmetric mode has twice the single string decay rate, because acting together the strings exert twice the force of one string on the bridge, the bridge moves further, and the decay rate doubles. Outside the mistuning range ± 1 there is only one decay rate, equal to the single string decay rate, the same as in 'traditional' theory.

If the hammer strikes both strings in the same direction, how can the antisymmetric mode be started? With a 'perfect' hammer strike and zero mistuning, all three strings would start moving in the same direction at the same time, and only the symmetric mode would result.

This would decay at twice the rate of the single string. Antisymmetric mode motion is introduced (a) through 'imperfections' in the hammer strike, and (b) through the presence of mistuning. In an 'imperfect' hammer strike, each string will not begin with exactly the same amplitude.

To begin with, both strings are 'driving' the bridge motion together. As the (first) string with least initial amplitude approaches zero amplitude after a short period of decay, the bridge which is still being driven by the other (second) string, begins to 'drive' this first string, causing it to move.

A string driven by a moving bridge, always begins to move 'out of phase' with the bridge driving it. The second string is *in phase* with the bridge motion because it is still driving it, rather than being driven by it. Thus motion builds up in the first string, out of phase with the second string, and the antisymmetric mode motion begins.

We need to combine mode frequencies and decay rates to arrive at *generic decay patterns*. In the hypothetical case of a purely resistive bridge, at close to zero mistuning we should expect not exactly a beatless linear decay, but rather, a 'single null' decay pattern, rather like one *prompt*-sound beat followed by a linear decay of *aftersound*. Fig. 12 shows two such decay curves together with regular beat pattern decay.

Fig 12

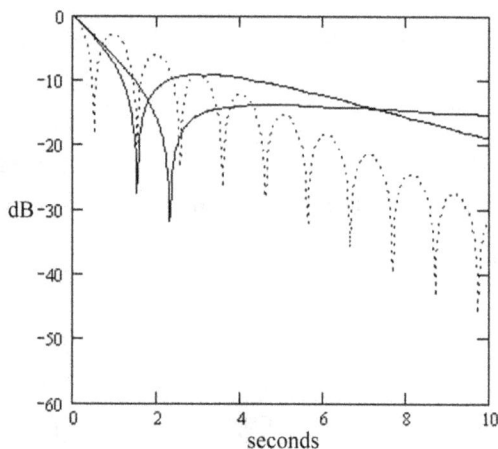

Decay curves (solid line) for a two string unison when the bridge is (hypothetically) purely resistive. The two solid line decays are for different mistunings – the change in mistuning affects not a beat rate, but the *decay rate* of the *aftersound*. The dotted line gives the standard 'heterodyne-like' pattern for comparison, that will occur at greater mistuning on the purely resistive bridge.

The decay behaviour in this hypothetical case of a purely resistive bridge changes abruptly from a 'regular beat' pattern to a 'single null' pattern when the mistuning becomes equal to the single string decay rate. The 'single null' generic decay pattern that occurs here within the mistuning range ± 1, comes from the hypothetical case of a purely resistive bridge, but should not be dismissed for this reason. As we shall see below, a similar 'single null' generic decay pattern occurs also as a prominent feature in the 'realistic' case of a resistive and reactive bridge, for a full trichord string group.

Resistive and reactive bridge

The dashed and dotted lines in Fig. 10 B are decay rates for a 'realistic' bridge that is both resistive and reactive. Fig. 11 shows the mode frequencies for the same bridge. In general now there are always two different mode frequencies and two different decay rates.

When there are two different mode frequencies, there is in principle a beat, but the beat will in general decay faster than the overall decay rate because there is also a difference in mode decay rates. (The beat amplitude can also *grow* as the overall decay progresses).

The net result of differing mode decay rates, in the case where the bridge has 'realistic' parameters, is a *dual decay rate* for the unison - a 'prompt sound' followed by an 'aftersound', which in the middle of the compass may have a ratio of around 8:1.

In this more realistic case of a bridge that is both resistive and reactive, the decay curves, as mistuning approaches zero, are more likely to be like those shown in Fig. 13.

Fig 13

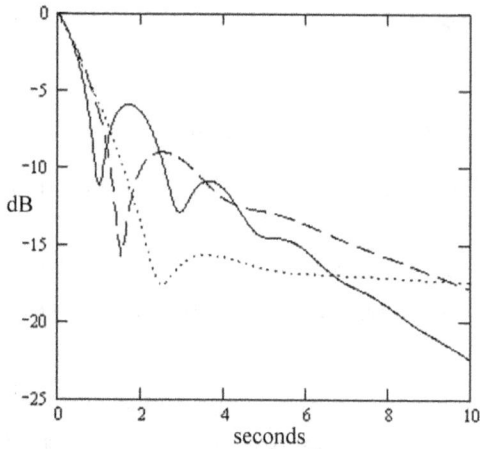

Decay curves from the Weinreich model when the bridge is both resistive and reactive. Mistuning is decreasing from the solid line to the dashed line to the dotted line.

We can see that in each curve the beat ends up decaying faster than the overall decay rate, so that finally, the decay becomes 'linear'. When the beat decay rate is sufficiently great, as in the dashed and dotted curves, only the first beat tends to be obvious, and the decay curve emulates the 'single null' of the purely resistive bridge (Fig. 24). The dual decay rate becomes more pronounced, the closer the mistuning is to zero.

Limitation of the Weinreich one plane model

It is possible to extend the Weinreich model to take into account a third string, and/or two planes of motion. For motion in one plane only, the matrix will have the same number of rows and columns as the number of strings. For a unison with either two or three strings, the matrix accounting for motion in *two* planes will have twice as many rows and columns as the matrix for one plane only.

Nevertheless, in the case where there is no anisotropy in the boundary admittance, the larger matrix for two planes will produce the same results as the smaller matrix for one plane. The larger matrix is in this case mathematically equivalent to the smaller one.

A difference in results from the one and two plane models arises, from the fact that the *resistive* (dissipative) part of the bridge admittance in the piano is *never* isotropic. One might think that the reactive ('spring' and 'mass-like') properties of the soundboard would always be anisotropic, because the main bridge-soundboard structure is designed in such a way as to have much greater 'give' in the direction perpendicular to the soundboard surface.

However, the bridge pin appears to lower the admittance in the direction parallel to the soundboard surface, making it possible for the admittance at the end of the string's speaking length to be isotropic.

Weinreich has shown that the reactive part of the admittance can be about the same as the resistive in the vertical direction (for a grand piano), whilst in the horizontal direction the reactive part remains the same as in the vertical, but the resistive part is about a quarter of its magnitude in the vertical.

This means that in the one plane model the reactive and resistive parts are set equal. In the two plane model, however, unless one assumes vertical mode axes, the reactive and resistive parts of the admittance cannot be set equal in the matrix, even when the reactive part is isotropic.

Fine tuning theory for unisons with false strings - two strings in two planes

Extending the Weinreich model

In the case where one or both of the strings in a unison pair are false at a particular partial j, the behaviour of the unison must be described in two planes. The admittance of the bridge must itself then be considered in two directions.

Mathematical description

Weinreich's description for the coupled system is comprised of two unison group strings vibrating in one plane,[13] and utilizes a dynamical matrix Ω containing all the essential information about the normalized system's configuration, necessary to determine its response patterns from the matrix eigenvalues. Each 'elemental' oscillator k with displacement q (both in the same plane) has the equation of motion

$$q_k = \text{Re}(\psi_k)$$

(3)

and the general equation of motion for the coupled system derived from the perturbation theory is

$$\frac{d\Psi}{dt} = i\Omega\Psi$$

(4)

where Ψ is a vector of all the Ψ_k s, and Ω is the dynamical matrix whose elements kk' are the frequency perturbation on oscillator k by the coupling mechanism and oscillator k'. For the two oscillator system the matrix Ω is:

$$\Omega = \begin{pmatrix} 2\varepsilon + \zeta & \zeta \\ \zeta & \zeta \end{pmatrix}$$

(5)

in which 2ε is the total 'mistuning' between the two strings, considered as applied entirely to one string only, and ζ is the frequency perturbation to one string caused by the bridge admittance and the other string, given by the relation

$$\zeta = i\omega_0 Z_0 Y/\pi$$

(6)

where ω_0 is the unperturbed partial frequency associated with a rigid bridge, Z_0 is the wave impedance (characteristic impedance) of the string for that partial, and Y is the bridge admittance. The perturbation has both 'resistive' and 'reactive' parts, causing both real frequency change and decay. Since ζ is complex it is written as $\xi + i\eta$ for a reactive value ξ and a resistive value η.

55

In the normalized model the real part of the bridge admittance ranges from −1 to +1. A positive value would correspond to a 'mass like' or inertial bridge property, which acts to raise frequencies. A negative value would correspond to an elastic bridge, which acts to lower frequencies. A zero real part would indicate no reactive component of the bridge − this corresponds to a rigid bridge. In practice the bridge has both inertial and elastic properties, but overall the frequency shift from ω_0 must be either positive or negative corresponding to a dominantly 'mass like' or 'spring like' bridge condition.

The imaginary part of ζ in this convention can only be positive, ranging from 0 to i, and corresponds to the damping effect of the bridge, with no appreciable change to the frequency ω_0. The decay rates are expressed relative to the single string partial decay rate, designating this as $\eta = 1$.

For string partials that are not perfectly in unison (the same frequency) the relative (angular frequency) 'mistuning' is designated 2ε as the perturbation due to 'mistuning', to the uncoupled angular frequency ω_0 of the strings.

The partial of string 2 is given the angular frequency ω_0, and that of string 1 is given the angular frequency $\omega_0 + 2\varepsilon$. (This follows Weinreich's use of a term 2ε rather than designating the whole 'mistuning' difference between the partials as ε. This is consistent with the approximation technique for producing Ω).

It is possible to construct a model representing a situation where motion in two planes is accounted for, and one or both strings have anisotropic localized boundary conditions at the bridge, leading to falseness.

In this case the uncoupled system has potentially four oscillators corresponding to the two pairs of individual string normal modes (it cannot be assumed, however, that in the general case that the normal modes of the coupled system have their mode axes aligned with those of the uncoupled system). Fig 14 represents the model adopted here for the arrangement of oscillators in the uncoupled system, for strings 1 and 2:

Fig 14

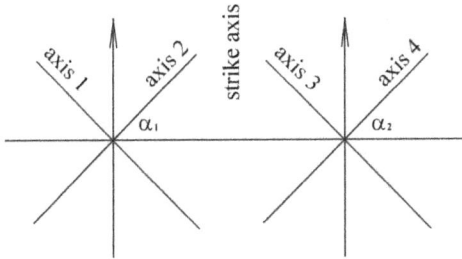

For simplicity the model is limited by setting $\alpha_1 = \alpha_2 = \pi/8$. The first condition to consider is when the system is without localized variations at the boundary of either string.

The bridge admittance in the model is consistent with the parameters already discussed, i.e. its reactive part is isotropic around the string's speaking length axis, whilst its resistive part is the same magnitude as the reactive in the vertical axis, but a quarter of this in the horizontal.

Because the uncoupled system's mode axes are aligned at 45 degrees to the horizontal and vertical, the resistive and reactive values of the bridge admittance for each of the uncoupled mode axes, can then be set as:

$\eta_1 = \eta_2 = \eta_3 = \eta_4 = 0.625$, for the resistive, and

$\xi_1 = \xi_2 = \xi_3 = \xi_4 = 1$, for the reactive.

Thus the bridge perturbation in each uncoupled axis direction can be written as $\zeta = 1 + 0.625i$. Localized variations for the reactive values in the directions of axes 1 and 3 (corresponding to the effects of the bridge pin) are then designated L_1 and L_3, and the matrix becomes

$$\Omega = \begin{pmatrix} \zeta + L_1 + 2\varepsilon & 0 & \zeta & 0 \\ 0 & \zeta + 2\varepsilon & 0 & \zeta \\ \zeta & 0 & \zeta + L_3 & 0 \\ 0 & \zeta & 0 & \zeta \end{pmatrix}$$

(7)

In Weinreich's one-plane description the coupled system exhibits two normal modes, one 'symmetric', and one 'antisymmetric', aligned with the conventionally chosen 'vertical' direction of the single string oscillators.

In the symmetric mode the two oscillators are 'in phase', and in the antisymmetric mode they are 'out of phase'. As is the case in Weinreich's model, the configuration of the present model for two planes yields normal modes that are *symmetric* and *antisymmetric*, but in this case now they are in orthogonal pairs.

Hypothetical resistive bridge

In the hypothetical 'test' case of a purely resistive bridge, both the real and imaginary parts of the eigenfrequencies of the 4 X 4 matrix Ω, appear in two corresponding pairs.

When there are no localized perturbations, the two pairs of curves for the real and imaginary parts of the eigenfrequencies - corresponding respectively to the coupled system's normal mode frequencies and their decay rates - each coincide.

When a localized perturbation is introduced, this has the effect of shifting one pair of eigenfrequencies (both the real and the imaginary) relative to both axes. Figs 18 and 19 show the effect of introducing a localized perturbation on axis 3, by putting $L_3 = 3$.

Fig 15

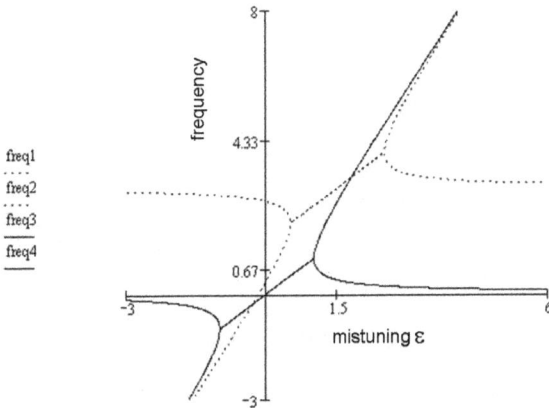

freq1
. . . .
freq2
. . . .
freq3

freq4

Real eigenfrequencies of Ω when $L_3 = 3$, for a purely resistive bridge.

Fig 16

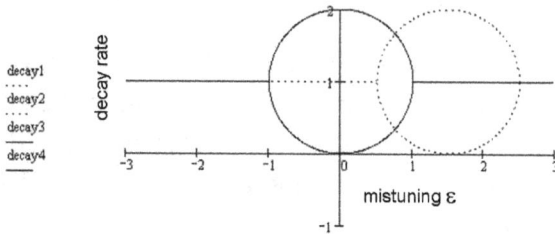

Decay rates of Ω when $L_3 = 3$, for a purely resistive bridge.

'Realistic' bridge

In the case of the 'realistic' bridge parameters with the reactive component present, the eigenfrequency pairs are offset by the presence of a single localized perturbation, Figs 17 and 18.

If $L_1 = L_3 \neq 0$ the decay rates (imaginary eigenfrequencies) then coincide, and the real eigenfrequency pairs are offset in frequency, Fig 19.

Fig 17

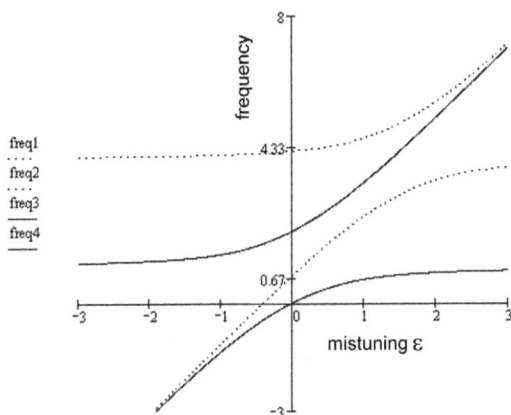

Real eigenfrequencies of Ω when $L_3 = 3$, for a reactive and resistive bridge.

Fig 18

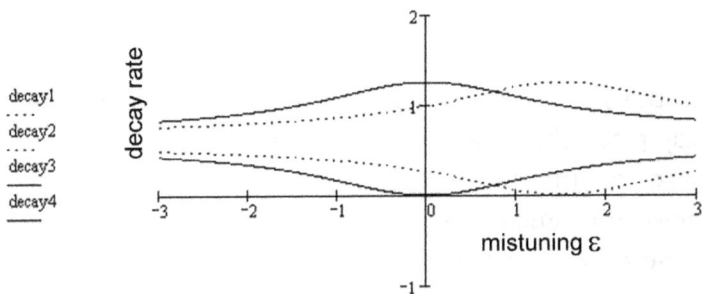

Decay rates of Ω when $L_3 = 3$, for a reactive and resistive bridge.

Fig 19

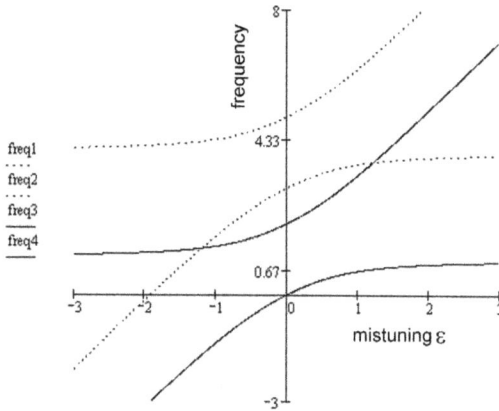

Real eigenfrequencies of Ω when $L_3 = L_1 = 3$, for a reactive and resistive bridge.

It is the generic decay curve *patterns* rather than specific decay rates, that are of interest. These are specifically with regard to the piano tuning principle of attempting to achieve the 'smoothest' or most 'near linear' decay curve for the audible partial.

The curves represent motion in line with the strike axis. Energy from horizontal string motion will be transmitted through vertical motion of the soundboard (motion perpendicular to the soundboard) if the string-bridge-soundboard system's normal mode axes are *not* vertical and horizontal.

Another way of looking at this is that vertical and horizontal motion in the bridge-soundboard structure is actually coupled, rather than the vertical and horizontal being normal mode directions for the coupled system (this is just an instance of the principle that a shift from an 'uncoupled' to 'coupled' description corresponds to a rotation of co-ordinates).

In the hypothetical example of a purely resistive bridge, at zero mistuning the antisymmetric modes have zero decay rates, and the symmetric modes have twice the single string decay rate in that axis direction.

However, an antisymmetric mode will not be present at zero mistuning unless there is a difference in the initial conditions of the two component oscillators. Weinreich suggests how 'imperfections' in the hammer could lead to the presence of the antisymmetric mode, from the initiation of the system.

The model used here to generate general motion decay curves of the partial (as a superposition of the coupled system's four normal modes at one partial number n), takes into account mistuning as the only factor that introduces the antisymmetric modes.

In the physical case 'imperfection' in the hammer strike, including strike misalignment, especially when there are grooves worn in the hammer face, could also contribute to antisymmetric mode excitement.

For the generic model the eigenfrequencies of $\Omega(\varepsilon)$ can be applied as direct perturbations to an otherwise equal set of complex mode frequencies. In such a model it is expedient to use the mode decay rates themselves as a suitable reference for the initial amplitudes of the modes, for a given mistuning.

Assuming no hammer irregularity, then in the example of a purely resistive bridge, at zero mistuning the antisymmetric decay rates must be zero and the initial antisymmetric amplitudes must consequently be set at zero. The symmetric decay rates, on the other hand, are twice the single string decay rate at zero mistuning.

It can be seen from the graph that at larger mistunings both the symmetric and antisymmetric decay rates tend towards unity (the single string decay rate) for modes 2 and 4, or 0.25 for modes 1 and 3.

Each normal mode p_n of the coupled system (where $n = 1$ to 4) is then written as

$$p_n(t,\varepsilon) = \text{Re}\left[\exp\left\{i\left(\omega_0 + v_{Rn}(\varepsilon) + iv_{In}(\varepsilon)\right)t\right\}\right]$$

(8)

where $v_{Rn}(\varepsilon)$ is the real part of eigenfrequency n and $v_{In}(\varepsilon)$ is its imaginary part. The unperturbed frequency ω_0 here can in principle be any value (since we are interested only in functional dependence on time and ε).

However, in the present model, the method employed to plot the resultant decay curves shown below, involves sampling the actual wave form produced in the vertical direction. For speed of computing considerations, ω_0 is included at a suitably high value (arbitrarily set at 440 Hz) compared to the sampling frequency.

The four normal modes must be superposed in two pairs to produce the overall general motion described in two orthogonal directions. Two orthogonal general motions, each superposed from one symmetric and one antisymmetric mode are possible, and these are written in the model as

$$G_1(t,\varepsilon) = p_1 v_{I1} + p_2 v_{I2}$$
$$G_2(t,\varepsilon) = p_3 v_{I3} + p_4 v_{I4}$$

(9)

in which the v decay rates serve to determine initial amplitudes, as proposed above. If the coupled system's normal mode axes are inclined to the 'vertical' antenna axis and 'strike' axis, then both general motions will contribute to the final motion in the 'vertical' antenna direction, perpendicular to the soundboard.

For unit amplitude in the 'strike' axis, the *relative* initial displacement amplitudes in the normal mode directions would depend on the directions of these modes, which in turn would be determined by the alignment of the uncoupled system's mode axes. However, this can be circumvented by setting

$$\alpha_1 = \alpha_2 = \pi/8.$$

(10)

Setting the α's values equal at 45 degrees provides equal initial amplitudes of the 'strike' in each mode axis direction. If only one string is 'false', then the model's uncoupled 'mode axis' for the non false string is in fact arbitrarily in line with the strike axis.

However, it is still legitimate to use axes set at 45 degrees to the vertical and horizontal to describe its motion. Thus the model finally produces a resultant wave form $C(t)$ in the 'strike' or vertical antenna axis direction, for any fixed mistuning value, given by

$$C(t) = A_1 G_1(t) \sin 45° + A_2 G_2(t) \cos 45°,$$

(11)

which for equal initial string displacements A_1 and A_2, means representations of the generic decay curve shapes can be plotted using simply

$$C(t) = G_1(t) + G_2(t).$$

(12)

Eqn 25 is a model for the situation in which $\alpha_1 = \alpha_2 = \pi/8$, rather than a general expression of bridge motion as a function of mistuning, α_1 and α_2. To generate a curve representing the overall decay pattern in the partial, $|C(t)|$ was sampled by integrating forward over 1/100th of a second (roughly 4 cycles) at each sample point, at 20 samples per second, for 5 seconds. This produces a set of samples E_n for 100, for the decay curve. Thus for unit unperturbed decay rate the graphs of y_n (in dB) are for:

$$y_n = 10 \left[\log\left(E_n \exp-\frac{n}{20} \right) - \log\left(E_1 \exp-\frac{1}{20} \right) \right]$$

(13)

where the horizontal axis n represents units of 1/20th second. The value of ε in $G_1(t,\varepsilon)$ and $G_2(t,\varepsilon)$ in the generation of all the graphs is given using $\varepsilon = \pi M$ where M is the mistuning in Hz,[14] for the model's theoretical single string partial j with unit decay rate.

Understanding the tuning behavior
of the two-string unison

When $L_1 = L_3 = 0$ corresponding to a unison in which neither string is false (for this partial number), sufficient 'mistuning' between the strings produces the expected decaying 'heterodyne beat like' pattern. Fig 20 shows the pattern for a mistuning of 1 Hz (producing a beat rate of 1 per second).

Fig 20

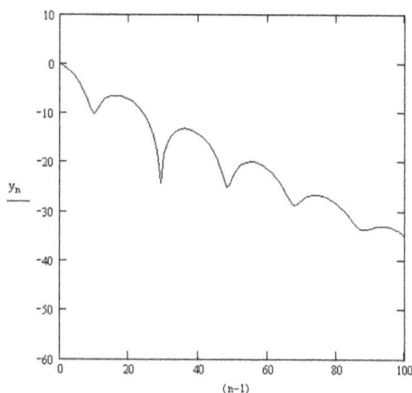

Theoretical decay pattern of one partial j for the unison when neither string is false, and there is a mistuning of 1 Hz at partial j.

It was stated above that the local variation in the reactive part of the bridge admittance necessary to produce a false beat rate of 1 Hz, would need to be in the order of 6:1. Putting L_1 or L_3 in eqn 6 equal to 6, with zero mistuning, produces the curve Fig 21. The beat *rate* from the false string is 'inherited' by the decay of the unison.

Fig 21

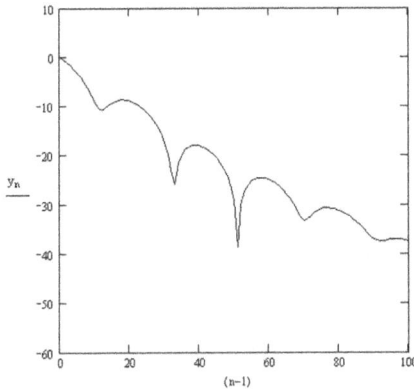

Theoretical decay pattern of one partial *j* for the unison when one string has a false beat rate of 1Hz at partial *j*, and there is no mistuning between the strings at partial *j*.

For a sufficiently small false beat rate, there is an 'optimum range' of mistuning in which the false beat rate is not obviously inherited by the unison decay. In this model the largest beat rate for which such an 'optimum range' can be found is approximately 0.7 Hz. This can be obtained by setting the perturbation L_1 or L_3 to 3 (Fig 22). With increasing mistuning, the decay pattern temporarily becomes *less* 'turbulent', and more 'quasi-linear', up to

69

mistuning values of around 0.3 Hz - 0.4 Hz (the equivalent mistuning of a beat rate of around I beat every two or three seconds, in the absence of any falseness). For mistunings greater than this, 'turbulence' in the decay begins to increase until at around I Hz mistuning, a beat rate equivalent to the mistuning begins to assert itself in addition to the false beat.

Fig 22

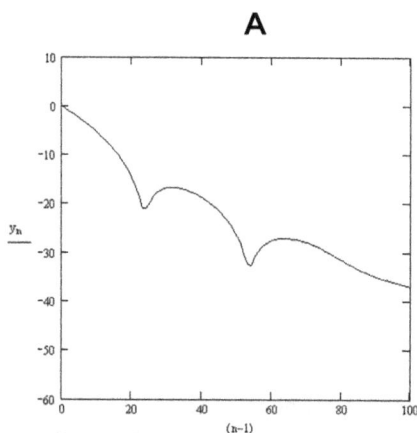

A

Theoretical decay pattern of one partial j for the unison when one string has a false beat rate of 0.7 Hz at partial j, and there is no mistuning between the strings at partial j.

B

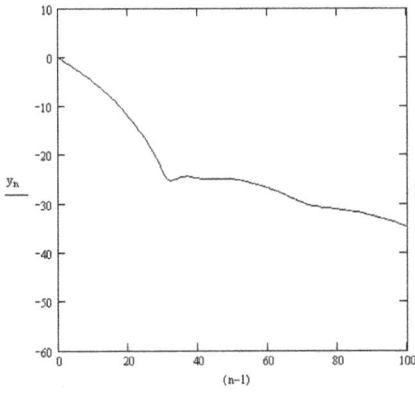

Mistuning increased to 0.2 Hz between
the strings at partial *j*.

C

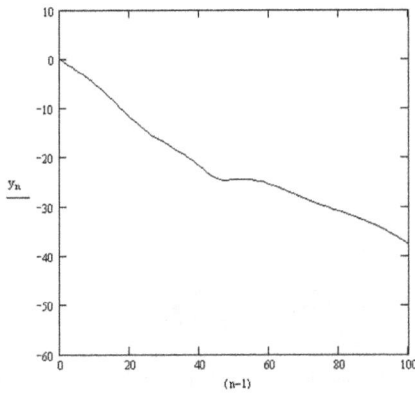

Mistuning increased to 0.3 Hz

D

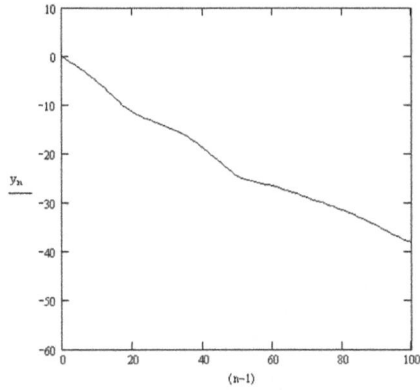

Mistuning increased to 0.4 Hz. This mistuning, or perhaps the preceding one, would be an optimum for a 'quasi-linear' decay.

E

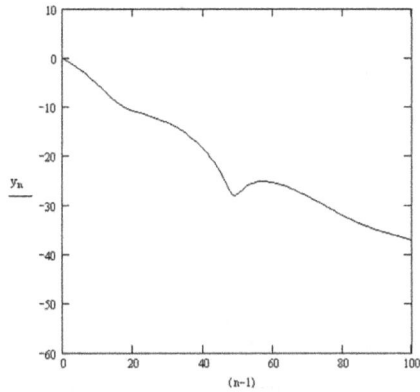

Mistuning increased to 0.5 Hz. The beat starts to reappear.

F

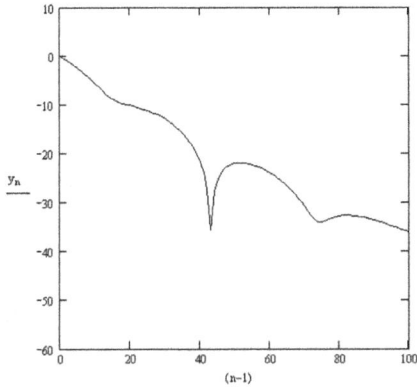

Mistuning increased to 0.6 Hz.

G

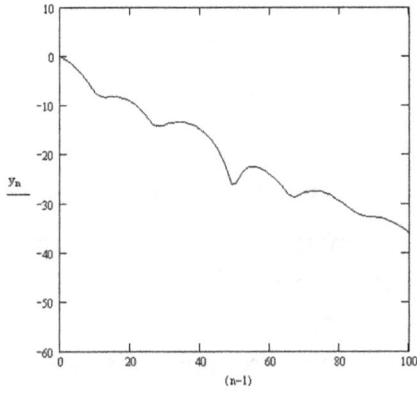

Mistuning increased to 1 Hz.

H

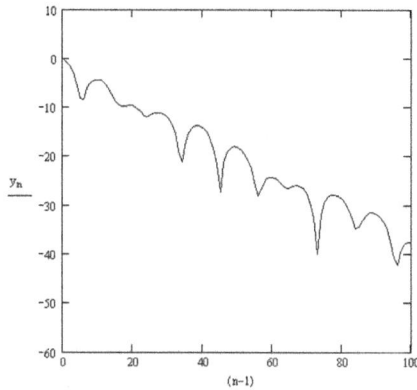

Mistuning increased to 2 Hz between the strings at partial *j*. The beat pattern is now a mixture of both the false beat and a beat due to the mistuning.

For the faster false beat rate ($L_3 = 6$) of 1 Hz no such position of quasi-linearity in the decay curve can be reached.

The curves generated for y_n are of course for a unison with specific parameters, but they nevertheless serve to illustrate the generic principle where there is the attempt to achieve the greatest possible 'linearity' or 'smoothness' in partial decay, or the attempt to 'eliminate' as much as possible from the decay curve, any 'beat-like' pattern.

Of course, this deals with a single partial. In actual tuning, there will be many partials contributing to the tone of the unison, that can simultaneously display this kind of behaviour. Some will be more prominent than others, and the tone of the unison will depend on the way they mix. The art of tuning the unison, as distinct from the science, lies in achieving the best aesthetic result, in this situation.

74

Trichord tuning

A trichord may be tuned in two string pairs, or the third string may be tuned with the first two tuned strings sounding (using no wedge). The behaviour of the full trichord will be of particular interest since it is the full trichord that sounds in normal playing, unless the *una corda* pedal is in use.

The trichord in two planes is represented using a 6 X 6 matrix. With similar axes configuration, the 6 X 6 matrix also allows mistuning between two strings as a way of preventing the trichord inheriting a false beat from a single string. Mistuning within two pairs of strings in the trichord can also sometimes provide a 'better' result than that obtainable from the 4 X 4 matrix for the same parameters.

In practical tuning terms, this corresponds to a situation in which under certain conditions, a false beat in one string that is not successfully hidden on tuning the second string, can be more successfully hidden on addition of the third string.

There is, in general, more opportunity in the trichord to alter the inheritance from a false string so that the resulting decay in the trichord does not resemble a regular beat, and irregularities or fluctuations in the decay curve are 'less prominent'.

This would be expected from the increased number of parameters. In the 4 X 4 matrix there are 8 non-zero elements whereas in the 6 × 6 matrix there are 18. There is of course also correspondingly more opportunity to *introduce* irregular fluctuations in the decay curve through mistuning.

In either the string pair or the trichord unison group the two-plane model becomes effectively redundant in the case where there is no false beat in any string. In such a case, for example, the 4 × 4 matrix used above, produces the same end results as the Weinreich 2 × 2 matrix (there are of course 4 eigenfrequencies rather than 2, but they coincide exactly in pairs, reducing in effect, to one pair as in Weinreich's results).

Use of a 3 × 3 matrix for the one-plane trichord therefore provides a good insight into the behaviour of the 'pure' trichord. In this case, the matrix becomes:

$$\Omega = \begin{pmatrix} \zeta + 2\varepsilon_1 & \zeta & \zeta \\ \zeta & \zeta & \zeta \\ \zeta & \zeta & \zeta + 2\varepsilon_2 \end{pmatrix}$$

$$(14)$$

The positioning of the mistunings ε_1 and ε_2 on the diagonal does not affect eigenfrequencies. There are of course 3 relative mistunings between the three strings, but the entire mistuning configuration can be specified relative to one string, as 2 mistunings.

It is useful to set $\varepsilon_2 = 0$ and examine the behaviour as a function of mistuning ε_1. The eigenfrequencies for the hypothetical purely resistive bridge, can again be used to ascertain the distinct, basic normal mode configuration and behaviour:

Fig. 23

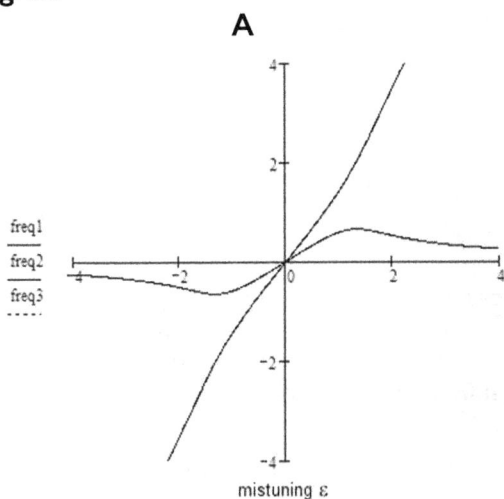

Real eigenfrequencies of the trichord for a purely resistive bridge, when two strings are at zero mistuning. The third eigenfrequency runs along the horizontal axis.

B

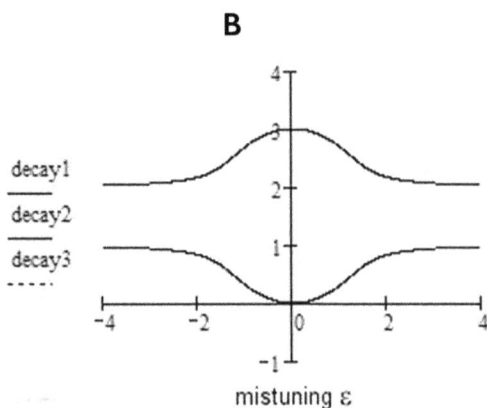

decay1
———
decay2
———
decay3
- - - -

mistuning ε

Decay rates (imaginary eigenfrequencies) of the trichord for the purely resistive bridge, when two strings are at zero mistuning. The third eigenfrequency runs along the horizontal axis.

Now we see that at zero mistuning the decay rate of one mode reaches 3 times the single string decay rate, whilst a second mode reaches zero decay rate. The third mode has zero decay rate and is independent of mistuning.

The decays are now permanently separated as they would be for a two string unison on a bridge with a reactive component. The real eigenfrequencies intersect at zero mistuning, and one eigenfrequency is permanently unperturbed by the mistuning. How does this translate to the physical situation?

Trichord modes

The coupled trichord system has three normal modes, which we can label modes A, B and C. Mode A is the only symmetric mode, and consists of all three strings moving together, in phase. Mode A thus has three times the decay rate of the single string, as represented by the upper solid line of Fig. 23 B, at zero mistuning.

Mode B is antisymmetric, and consists of two strings moving 'symmetrically in phase' whilst the third string motion is antisymmetric to, and out of phase with, the motion of other two. At zero mistuning its decay rate appears on the lower solid line of Fig. 23 B.

In mode C two strings are in antisymmetric motion, whilst the third, like the bridge itself, does not move. This is also an antisymmetric mode. The frequency and decay rate of this mode are independent of mistuning, and follow the horizontal axes in the graphs.

General motion of the system can be a superposition of all three modes, but as in the case of the unison string pair, a 'perfect' hammers strike will initiate only the symmetric mode A. If the initial motion contains some of mode B, due to mistuning or 'imperfection' in the hammer strike, then in principle mode A can decay into mode B, which in turn can decay into mode C.

When the *una corda* pedal is in use, mode B will be initiated by the strike, because the third string will not be struck, and will be driven into motion by the bridge, which itself is driven by the other two strings. For the *una corda* strike, motion begins symmetrically between the two struck strings, but antisymmetric motion with the third string will begin to grow immediately.

In effect, the motion is practically antisymmetric from the beginning, providing a more sustained tone for quiet playing, than the full trichord. Furthermore, as decay progresses, any small antisymmetric component that was initially present between the two struck strings, due to mistuning or hammer 'imperfection', can cause the initial mode merely to degenerate into another mode B form, or mode C, and so on.

For example, if there is an initial difference in amplitude between the two stuck strings, the first string to approach zero amplitude may remain at zero amplitude whilst the other two strings remain in antisymmetric motion (mode C). Depending on the relative amplitudes, this first decaying string may alternatively be driven into motion in phase with the unstruck string (mode B).

Whilst there is only one form of symmetric mode A, each of the other two antisymmetric modes have three possible alternative degenerate forms, depending on which string plays which role in the mode motion.

Conclusions

Partials, inheritance, beats and decay rates

In 'traditional' piano tuning theory heterodyne beats occurring in the tuning of two unison strings, are created by the interference of two partials, one from each string. From this concept it is assumed that the beat rate in at any partial number, is directly proportional to the mistuning between the strings.

Modern acoustical physics, which treats the strings and soundboard/bridge as a complete system, reveals that this is not the case. The 'traditional' idea of two interfering partials, depends on the notion that the spectrum of partials for each individual string, persists when both strings are sounding.

In modern acoustics, when both strings are sounding together, a new system exists, consisting of two strings plus the bridge, which has its own spectrum of partials. Many of these partials in the new spectrum do correlate to the partials that would be found in the individual string spectrums. However, in the finest tuning, those that do not, are precisely those that contain the movement or beats to which the tuner listens.

Individual strings each have their own spectrum of partials, but a unison group of strings has its own spectrum of partials, that *inherits* some of the properties from the single string spectrums. In the finest unison tuning, the unison's partials may always contain a 'beat' even when mistuning between the strings is zero, but the 'beat rate' of this beat may be so slow compared to the decay rate, that the partial appears to be 'beatless'. There is a range, or *coupling region* in which tiny changes of tension on one string do not significantly alter the beat rate in a partial, but rather, alter the decay rate of that partial.

The 'soundscape'

The spectrum consists of a large number of partials. A change to string tension generally affects the spectrum as a whole. Concentration on the behaviour on of one partial, whilst convenient in theoretical investigation, can be misleading in that more than one partial is involved in determining the qualities and properties of the unison sound.

The spectrum *as a whole* must therefore always be considered in describing what the tuner listens to. The perceived sound of the spectrum as a whole, when it is not being perceived as a set of discreet individual partials, but *is* being perceived in terms of *movement* or beating in the overall decay pattern, can be described as a *soundscape*. Thus, it can be said that whilst the musician hears the tone of the unison, the trained piano tuner listens also to its *soundscape*.

'Significant partials'

The set of all partials present in the unison spectrum, may have a subset of partials (which may also be equal to the set of all partials present) that can be referred to as *significant partials* for tuning purposes. These can be proposed to be the partials that most critically affect the sound quality of the unison.

What makes a partial 'significant' in this context, may depend on its decay rate, its amplitude at any time, and its position in the spectrum. But it may also depend on subjective factors that fall within the field of psycho-acoustics.

The concept of *significant partials* is nevertheless useful when talking about partials, in order to make it clear that certain statements do not necessarily apply to *all* partials, but only to those which will have a marked effect on the perceived quality of the soundscape.

Partials and beating – differences between 'traditional' and contemporary theory

'Traditional' theory considers partials to be normal modes of the string. Contemporary theory distinguishes between partials and normal modes. In contemporary theory two or more normal modes may constitute one partial.

Contemporary analysis challenges the 'traditional' view that in order to eliminate beating at a partial number j, the piano tuner is 'matching the frequencies' of partials from the individual strings. In the contemporary view, in the act of eliminating the beat at partial j in the unison with more than one string sounding, the jth partials from the individual strings are no longer present as such.

There are no individual string partials there, to be 'matched'. Rather, by adjusting the tension in one string, one is altering the two normal modes responsible for the partial j (which are not individual string normal modes), and in doing so is altering the beat rate and/or the decay rate of the unison's partial.

The inheritance of false beats

A single string may have one or more partials in its spectrum that beats as it decays. Generally this may be due to anisotropy in the reactive part of the bridge admittance, but there can be other causes.

When the string is a member of a unison group, the spectrum of the unison group (when both or all three strings are sounding) may *inherit* a beat in the decay of a partial j, if a beat was present in the decay of partial j of the single string sounding on its own.

It is not, however, the 'false beat' in the single string that is being heard, when the unison group is sounding. The inherited beat in the unison is not, so to speak, the single string beat. The unison's spectrum must be thought of as an entirely new spectrum in its own right.

Unlike a partial of a single string, a partial of the unison's spectrum can be changed and controlled in ways not possible in the case of the single string. This is because in the unison, *mistuning* between the strings exists as a parameter of the system that can be changed, by changing the tension on one of the strings.

In the system of the single string, there is no mistuning parameter, only the string tension parameter with which adjustment can be made. The situation is rather like having an extra control built into the unison but not into the single string.

When neither string is false at partial j in its own spectrum, then adjusting one string in the unison group with the strings sounding together, enables the beat rate in the jth partial of the unison's spectrum to be finely controlled, or inside the *coupling region* it enables the decay rate of that partial to be controlled.

Hiding false beats

The two-plane model is a scientific illustration of Braid White's notion that false beats can be hidden through mistuning between the stings of the unison. It is still important to appreciate, however, that the situation is not what it might seem.

When a false beat is inherited from a single string by the unison pair or group of strings, we are not dealing with a system in which the 'false' string is autonomously producing the false beat, whilst the others are not.

It is indeed the *physical parameters* – e.g. its locally anisotropic boundary admittance - associated with that individual string, that are responsible for the unison producing a false beat. But once the string is sounding together with the others, it is the entire group that is responsible for producing the false beat that is heard.

By judicious introduction of mistuning between the strings, the way the entire group behaves is altered, and the false beat may sometimes be hidden. In general, the false beat needs to be sufficiently slow compared to the decay rate for this to be successful.

Two false strings

When both strings of a unison pair are false at partial j, then in general, the model in the configuration shown, fails to provide an opportunity to significantly hide the false beat at any mistuning. However, the model has been simplified by being made symmetrical.

Empirically, we know that false beat hiding is not entirely excluded as a possibility when both strings are false. For example, expert tuners will be aware from experience that where two strings of a unison exhibit sufficiently slow false beating at the same false beat rate, effective hiding of the false beat is invariably possible.

It should always be remembered that even though this may be the case, it is *all* the significant partials that count in the tuning of the unison, and not just a particular partial j. The final mistuning in the fundamental will depend on the behaviour of all significant partials, more than one of which may be false.

Appendix -
Some further points in theory

The unison system for one partial is modelled using an m X m matrix where m is the number of normal modes in the uncoupled system. This yields m complex eigenfrequencies for the coupled system. The matrix elements P_{mn} are the perturbation on oscillator m by the bridge coupling and oscillator n. Perturbations P_{12}, P_{21}, P_{34} and P_{43} are by definition zero. When just one string is allowed to sound, a false beat can be audible if it is mapped from both normal modes of a pair at partial number j, and this mapping can be represented by measuring the normal mode motions relative to 'antennae' axes, rotated relative to the normal mode axes. The eigenfrequencies of the model for the coupled system with two strings are found from a matrix in the form:

$$\Omega = \begin{pmatrix} P_{11} & 0 & P_{13} & P_{14} \\ 0 & P_{22} & P_{23} & P_{24} \\ P_{31} & P_{32} & P_{33} & 0 \\ P_{41} & P_{42} & 0 & P_{44} \end{pmatrix}$$

15

Eqn. 15 will expand by the top row to produce

$$\Omega = P_{11}\begin{pmatrix} P_{22} & P_{23} & P_{24} \\ P_{32} & P_{33} & 0 \\ P_{42} & 0 & P_{44} \end{pmatrix} + P_{13}\begin{pmatrix} 0 & P_{22} & P_{24} \\ P_{31} & P_{32} & 0 \\ P_{41} & P_{42} & P_{44} \end{pmatrix} - P_{14}\begin{pmatrix} 0 & P_{22} & P_{23} \\ P_{31} & P_{32} & P_{33} \\ P_{41} & P_{42} & 0 \end{pmatrix}$$

16

and expanding by top rows again we obtain

$$\Omega = P_{11}\left[P_{22}\begin{pmatrix} P_{33} & 0 \\ 0 & P_{44} \end{pmatrix} - P_{23}\begin{pmatrix} P_{32} & 0 \\ P_{42} & P_{44} \end{pmatrix} + P_{24}\begin{pmatrix} P_{32} & P_{33} \\ P_{42} & 0 \end{pmatrix} \right]$$

$$+ P_{13}\left[- P_{22}\begin{pmatrix} P_{31} & 0 \\ P_{41} & P_{42} \end{pmatrix} + P_{24}\begin{pmatrix} P_{31} & P_{32} \\ P_{41} & P_{42} \end{pmatrix} \right]$$

$$+ P_{14}\left[P_{22}\begin{pmatrix} P_{31} & P_{33} \\ P_{41} & 0 \end{pmatrix} - P_{23}\begin{pmatrix} P_{31} & P_{32} \\ P_{41} & P_{42} \end{pmatrix} \right]$$

17

The single isolated mode co-ordinate of string 2 in the case where only string 1 is false, is a scalar, and the oscillator's 'mode axis' will be determined by the vector of the strike axis. It can nevertheless be included in the model as two oscillators on axes in line with axes 1 and 2. With axes 3 and 4 aligned with 1 and 2, we can put

$$P_{14} = P_{41} = P_{23} = P_{32} = 0 .$$

18

Also, at zero mistuning we can consider all the localised perturbation as being on axis 1, and consider all perturbations as relative to

$$P_{22} = 0.$$

Then, also

$$P_{13} = P_{33} = P_{31}$$

and

$$P_{24} = P_{42}$$

We then obtain from eqn. 17:

$$\Omega = P_{11}P_{24}\begin{pmatrix} 0 & P_{13} \\ P_{24} & 0 \end{pmatrix} + P_{13}P_{24}\begin{pmatrix} P_{13} & 0 \\ 0 & P_{24} \end{pmatrix}$$

$$= \begin{pmatrix} P_{13} & P_{13} \\ P_{11} & P_{13} \end{pmatrix}\begin{pmatrix} P_{24} & 0 \\ 0 & P_{24} \end{pmatrix}$$

as the basic configuration for a two string unison with one false string, at zero mistuning. If there is no localised perturbation, then $P_{11} = P_{13}$, $\Omega = 0$, and no beat will result in the decay. The values of P_{13} and P_{24} will depend on the matrix for the bridge admittance – they may differ even for a bridge that is reactively isotropic, because the resistive part of the admittance is not isotropic. The eigenfrequencies of Ω in Eqn. 21 differ when $P_{11} \neq 0$, producing a beat whose beat frequency is a non linear function of P_{11}.

Notes

[1] J. Cree Fischer, *Piano Tuning / Regulating and Repairing*, (1907), reprinted as *Piano tuning. a simple and accurate method for amateurs* (Dover, NY, 1975).

[2] W. Braid White, *Piano Tuning and Allied Arts*, (1917, 14th reprint 1972), p. 106.

[3] The normal mode with the lowest frequency is known as the *fundamental*, and for a fundamental v, a harmonic series of frequencies would be v, $2v$, $3v$...nv, where n is a label called the *harmonic number*. Thus, for example, in a true harmonic series, harmonic number 5 would have 5 times the frequency of the fundamental. In the case of the piano string's motion, we would label its normal modes with the equivalent *mode numbers*.

[4] The 'imaginary number' i is the square root of minus one. A complex number is in the form $a + ib$, where a is the real part and ib is the imaginary part. The product of two positive imaginary parts will be real and negative. A *complex frequency* has both a real and imaginary part.

[5] Transverse (see note 8) waves can travel along the string and are partially or totally reflected at the boundaries. As a result, waves are moving in both directions along the string. These waves interfere in such a way that wave motion appears on the string, which stationary along the length of the string, but causes the string to vibrate transversely, in wave crests and troughs. These are called 'standing waves'.

[6] See note 3, above.

[7] See P. M. Morse, *Vibration and Sound* (New York, 1948), Sec. III, p. 13.

[8] H. Tanaka, K Mitzutani and K. Nagai, 'Experimental analysis of two-dimensional vibration of a piano string measured with an optical device', *J. Acoust. Soc. Am.* 105, 2, p. 1181 (1999), and 'Two-dimensional motion of a single piano string', *J. Acoust. Soc. Am.* 100, 4, p. 2843 (1996).

[9] Hundley, TC; Benioff, H; Martin, *op. cit.*, p. 1306.

[10] G. Weinreich, "Coupled piano strings", *J. Acoust. Soc. Am.* 62, pp. 1474-1484, (1977), p. 1479.

[11] G. Weinreich, (1977), p. 1480.

[12] G. Weinreich, (1977), p. 1481.

[13] Weinreich, *op. cit.*

[14] See Weinreich, *op. cit.*, p. 1475.